Esther Magambo

Sustainability of Group Based Fruit Processing Enterprises in Kenya

Esther Magambo

Sustainability of Group Based Fruit Processing Enterprises in Kenya

The Case of Gatundu and Thika Districts

LAP LAMBERT Academic Publishing

Impressum / Imprint
Bibliografische Information der Deutschen Nationalbibliothek: Die Deutsche Nationalbibliothek verzeichnet diese Publikation in der Deutschen Nationalbibliografie; detaillierte bibliografische Daten sind im Internet über http://dnb.d-nb.de abrufbar.

Bibliographic information published by the Deutsche Nationalbibliothek: The Deutsche Nationalbibliothek lists this publication in the Deutsche Nationalbibliografie; detailed bibliographic data are available in the Internet at http://dnb.d-nb.de.

Coverbild / Cover image: www.ingimage.com

Verlag / Publisher:
LAP LAMBERT Academic Publishing
ist ein Imprint der / is a trademark of
AV Akademikerverlag GmbH & Co. KG
Heinrich-Böcking-Str. 6-8, 66121 Saarbrücken, Deutschland / Germany
Email: info@lap-publishing.com

Herstellung: siehe letzte Seite /
Printed at: see last page
ISBN: 978-3-659-14464-6

Zugl. / Approved by: Nairobi, University of Nairobi, Dissertation, 2009

FACTORS IMPEDING THE SUSTAINABILITY OF FRUIT PROCESSING ENTERPRISES OWNED BY FARMER GROUPS IN THIKA AND GATUNDU DISTRICTS

DEDICATION

This work is dedicated to my children ***Carol***, ***Martin, Mercy*** and ***Brandon***

ACKNOWLEDGEMENT

A number of institutions and individuals were of great assistance to me during the study. First, I wish to most sincerely acknowledge the encouragement and support from my supervisor, Prof. Pokhariyal for his step by step guidance through the whole study process.

Secondly, I would wish to thank the Chairman of the Department of Extra mural Studies, Dr. Christopher Gakuu on behalf of the department for facilitating this research. In addition I do appreciate the other members of the academic staff of the department who assisted me in one way or another.

The inspiration from my classmates is highly appreciated. We shared a lot of knowledge and information together that finally enabled me to put this report together. I cannot forget the valuable input from the District Agriculture Office Thika. I wish to specifically acknowledge the support given to the study team by the District Agricultural Officer, Rebecca Githaiga. I am particularly indebted to Esther Wakala for her guidance during the field data collection. On the same note I must thank my research assistants led by Peter Adipo for their contribution in this study.

There are many other institutions and individuals who supported me during the study in one way or the other, and I have not mentioned their names in this report. Please accept my gratitude. May the almighty LORD, BLESS YOU.

However, take full responsibility for any errors and mistakes that may be contained in this report.

LIST OF TABLES AND FIGURES

ACRONYMS

ATIRI -	Agricultural Technology Input Response Initiative
BDS -	Business Development Services
CIGs -	Common Interest Groups
EEC -	European Economic commission
EED -	European Economic Development
FTC -	Farmers' Training Centres
GDP -	Gross Domestic Product
GOK -	Government of Kenya
HCDA -	Horticultural Crops Development Authority
IPAR -	Institute of Policy Analysis and Research
KAPP -	Kenya Agricultural Productivity Project
KARF -	Kenya Agricultural Research Foundation
KARI -	Kenya Agricultural Research Institute
MCSS -	Ministry of Culture and Social Services
MDGs -	Millennium Development Goals
MFA -	Mangoes From Above
MOA -	Ministry of Agriculture
NALEP-	National Agriculture and Livestock Extension Programme
NGOs -	Non-Governmental Organizations
NMK -	Njaa Marufuku Kenya
RATC -	Rehabilitation of Agricultural Training Centres
SACDEP-	Sustainable Agricultural Community Development Project
SIDA -	Swedish International Development Agency
SPSS -	Statistical Package for Social Sciences
UNDP -	United Nations Development Programme

ABSTRACT

The main purpose of the study was to identify the factors that hinder the sustainability of farmer group-owned fruit processing enterprises in Thika and Gatundu districts. Six of the enterprises constituted the sample for the study. Data was collected from a sample of group members including officials, extension officers from Ministry of Agriculture (MOA) and an official from Sustainable Agricultural Community Development Project (SACDEP).

It emerged from the study that five of the six farmer group-owned enterprises have been in existence for a long time and therefore are likely to continue operating. All the enterprises have been receiving support from the MOA mainly in form of capacity building. SACDEP, the other major supporter of some of the enterprises has provided technical and entrepreneurial skills training and fruit processing equipment. Five groups have embraced a democratic form of leadership comprising well constituted management committees which have ensured group cohesion and continuity. The sixth one, Kagaa's collapse as a group enterprise resulted from lack of democratic leadership.

The study revealed that insufficient and unreliable rainfall, high cost of farm inputs, diseases and pests have reduced fruit production in all the farmer group-owned enterprises, hence low volume of fruit processed products. Fruit processing is also constrained by lack of automated processing equipment in some enterprises while in others equipment broke down frequently. Marketing of processed products was mostly restricted to the local consumers in the rural setting because transportation and proper storage facilities were inadequate. Capital inputs and the inability to source for external financial support were found to be crippling constraints to enterprise growth of all the group-owned enterprises. The groups were unable to utilize the internal and external markets that they were linked to by SACDEP due to the above mentioned reasons.

In order to enhance sustainability of farmer group-owned enterprises in Thika and Gatundu districts, it is recommended that farmers get the necessary support to improve fruit production. It is further recommended that fruit processing is centralized so that it is run professionally and is able to meet market requirements. This may either be done by the farmers groups in a larger legal entity or by a private entrepreneur. The key to sustainability will be to de-link management of processing with ownership in either of the scenarios.

CHAPTER ONE: INTRODUCTION

This chapter presents the background relevant to the issue of value addition to fruits by enterprises owned by farmer groups. It describes the research problem, general and specific objectives, research questions and the significance and justification of the study. Also included in the chapter are the scopes of the study (limitations and delimitations) assumptions, definitions of important terms and the conceptual framework for the study.

1.1 Background

At the global setting, human development activities are mostly guided by the Millennium Development Goals (MDGs). In developing countries, particularly those from the Sub-Saharan region, the most critical of the MDGs relates to 'poverty reduction through employment creation and local economic development' (UNDP, 2006).

In Kenya, the Ministry of Agriculture (MOA) has the mandate to promote and facilitate production of food and agricultural raw materials for food security for all, employment creation, income generation and overall poverty reduction in Kenya (MOA, 2008). The MOA's strategic plan 2006 – 2010, underscores the importance of reviving agricultural institutions through which empowerment of farmers can be achieved. The MOA's ministerial brief notes; "we believe that it is through farmers empowerment that we can achieve our vision of being the leading agent towards the achievement of food security for all, increasing employment, income generation and poverty reduction." (Daily Nation, Wed 20th August, 2008 p 40)

In order to achieve the set targets, the MOA, through its various departments and parastatal corporations continues to work with farmers, development partners and stake holders to address among other challenges; improved farming practices, value addition for improved returns, marketing and development of an enabling legal and policy environment for increasing agricultural productivity. In this regard the MOA is currently undertaking the implementation of such projects as the National Agriculture and Livestock Extension Programme (NALEP2) funded by the Swedish International Development Agency (SIDA) covering 62 districts. Kenya Agricultural Productivity Project (KAPP) funded by the World Bank; Mangoes From Above (MFA) covering Machakos, Thika, Kirinyaga and Embu districts funded by a German organization EED (European Economic Development). The focus is on equipping small scale farmers with skills for improved crop production, processing, marketing and initiating savings and credit schemes for members by organizing them into self-help groups.

The MOA's extension services department has over the years in conjunction with other stakeholders and development partners made efforts to promote value addition on agricultural produce among small-scale farmers. These are often conducted as follow-ups on promotion of a certain enterprise in a given area. When production levels go high, usually farmers are not able to market their produce at premium prices. Such gluts in supply, which are seasonal for some perishable produce, then call for value addition for preservation and raising market prices. There are also situations where home economics officers train women groups on cooking and basic processing of a variety of foods and farm produce. The objective of such training is mainly for food preservation and consumption. The skills gained from this form of training are used later to initiate group ventures that are up scaled at some point to enter the larger local market. Value addition on farm produce is promoted at group level. These are self –help groups registered with the Ministry of Culture and social services (MCSS)

The groups are mainly of two types, on one hand the extension officers may begin with an already existing self help group; especially the women groups constituted for a different purpose other than value addition. On the other hand the value addition idea is sold to the farmers and a group is formed specifically to engage in value addition. In the third scenario, the farmers were mobilized in large numbers for specific extension activities such as soil and water conservation programme, in their respective areas and enterprises were promoted alongside the conservation aspect. These activities gave rise to groups whose objective was to promote their members farming ventures as organized groups.

The NALEP is one such programme born out of the Soil and Water Conservation Programme under SIDA. Under NALEP, farmers are mobilized to form common interest Groups (CIGs). These are groups of farmers who are interested in taking up certain agro-based enterprises. The extension officers in conjunction with other stakeholders take the group members through training to equip them with necessary skills for production, processing, marketing and business management. Some groups may start production and

processing. Others may start as marketing groups to market the produce they already have. There are those who start with production and eventually graduate to value addition and then marketing. However, the most common category consists of groups which start with marketing the raw products and gradually scale up their activities to value addition.
Some of the obvious factors that would limit the farmers groups as far as value addition is concerned are initial capital for the purchase of essential equipment, the procurement of operations' premises and the technical know-how. For the groups in Gatundu and Thika district (The former Thika District) which is the geographical area of study, there are cases where other stakeholders and development partners in the agricultural sector have provided support in getting equipment for the groups. SACDEP is one such organization which is very active in the region. The support has ranged right from providing all the required equipment on grant, up to requiring the group to contribute 20% of the total cost of equipment.

Usually the groups provide premises for operations. These also range from free access to community or public owned premises to rented premises. In rare cases, some groups gradually manage to own their premises. Kigotho (2008) observes that income and especially Agro income for small scale farmers remain elusive to them despite the great efforts they extend to production. Over the years, more emphasis has been placed on production with little attention to marketing. Consequently, farmers have become disillusioned because they end up with little or no income to reward their efforts in the farms. When marketing take place, it is often through cooperatives or societies and producers have no direct control over the pricing or the choice of the market.

SACDEP-Kenya, in conjunction with other development partners operating in the former Thika District region, has taken up the challenge to assist the farmer out of this hopeless quagmire by educating them on how to raise benefits from their produce through value addition. This particular programme trains farmers on how to add value to their produce through Agro-processing, preservation, packaging and marketing their produce. The initiative is in line with GoK′s policy of making agriculture a profit oriented undertaking. The communities working in CIGs are adding value to their fruits by processing them into juices, juice concentrates and dried fruit snacks.

GOK's commitment to development of agro micro enterprises is well documented. It was recognized as a key strategy in 1986 report on Economic Management for renewed Growth. Subsequent policy reports and Sessional Papers (1989, and 2002) targeting development in line with national goals of fostering growth, employment creation, income generation, poverty reduction and industrialization.

1.2 Statement of the problem

The foregoing background has attempted to highlight the effort of GoK together with her development partners particularly the NGOs towards the economic improvement of small scale farmers. Since the introduction of the C1Gs by SIDA and their re-orientation to agricultural production, marketing and lately value addition, heavy investment has been expended to ensure that these community farmers groups are self-sustaining and viable.

Focusing on groups dealing with value addition to fruits in Thika and Gatundu districts, it is clear that the agri-businesses have not been very successful. Despite the concerted efforts of the various players in assisting in groups' formation, production, processing and marketing of farm produce, sustainability remains elusive. The statistics held by the MOA headquarters show that the agribusiness ventures have not been very successful. The situation is quite perturbing because field reports reaching the MOA's Agribusiness Development Section indicate that some groups have simply abandoned what started as very profitable ventures. The worst cases are where equipment that was donated to the group by supporting organization is lying idle. Ministry officials concerned have even explored the mechanism of withdrawing donated equipment from the failed group ventures and giving them to others. However, the main question that remains unanswered is: "Why did the ventures fail and what guarantee is there that these other groups will succeed?" The rate of failure among the group owned agribusiness is reportedly high. The researcher is not aware of any research that has been conducted by the MOA focusing the underlying factors causing the failures of the community group owned value addition agribusiness. This study, therefore, set out to determine the factors that hinder the sustainability of community group owned agribusinesses, focusing on fruit processing agribusiness ventures.

1.3 Research Objectives

The overall objective of the study was to identify the factors that hinder the sustainability of fruit processing agribusinesses owned by community groups in Thika and Gatundu districts. Specific objectives of the study were to:-

1. Determine the factors related to group functioning dynamics that hinder the operations of agri-businesses owned by community groups.
2. Determine the critical factors regarding business management that hinder the sustainability of the fruit processing agribusinesses owned by community groups.
3. Identify interventions that would make fruit processing enterprises owned by community groups viable and sustainable.

1.4 Research Questions

1. Does the method of constituting the initial group affect the sustainability of the group's business?
2. Which are the key group leadership issues that threaten the cohesiveness of the group?
3. Which factors hinder member' commitment to group activities.
4. What constraints does the group face with regard to:
   i. Production?
   ii. Processing?
   iii. Marketing their products?
5. What external assistance has been provided to ensure the groups are self-sustaining?
6. What more assistance can be provided to improve agribusiness?

1.5 Significance and justification

The programme of value addition takes cognizance of the fact that there is radical change in business environment. The MOA as the responsible arm of GOK together with their development partners need to configure their service delivery to planning with their clients. For this reason the study was considered important because it was expected to reveal the underlying causes of failure for a programme which was well planned and executed with the best intentions. The negative factors could be resolved for projects which are underway. The causes of failure could be factored into the planning and implementation of future projects including roll outs.

Therefore, the outcome of the study is expected to benefit MOA officials working with the groups, the development partners who support the groups and the business development service providers who may want to replicate the programme in other areas. Most importantly, the findings of the study may be shared with the members of the groups so that they may appreciate the shortcomings of their agri-businesses with a view to improving them.

1.6 Scope of the study

This study focused on community farmer groups who are engaged in fruit processing in Thika and Gatundu districts in central province of Kenya. These were self-help groups whose activities of adding value to farm produce included fruit processing.

1.7 Assumptions

The main assumption in this study was that the adopted sampling method would generate a range of farmer groups that was representative in performance levels. These levels were: the completely failed, those struggling to service and those whose performance was considered average. It was also assumed that the group leaders, ordinary group members, agricultural extensive officers and officials of development support organization included in the study would respond honestly to the interview questions.

1.8 Limitations

The size of the sample of enterprises was rather small. However, the population from which it was drawn incorporated a whole range of enterprises – those that have failed, those that were struggling and those whose performance was average. Other participants in the study included all the following stakeholders: ordinary group members and officials and support service provided from MOA and SACDEP.

1.9 Definition of Terms

BDS: Any non-financial service to business offered either formally or informally. This definition goes beyond the traditional definition of training, consulting, information, marketing, technology and administrative services targeting micro and small enterprises (David, 2002).

Donors: Funding agencies that pay for development activities of a project, a programme or an organization.

Marketing: A set of arrangements by which buyers and sellers are brought into contact to exchange goods or services – the interaction of demand and supply.

Sustainability: The capacity that ensures that benefits continue beyond the period of intervention.

Intervention: The temporary, facilitative mechanism by which donors and facilitators try to effect change for the better.

Value Addition: Make basic changes to raw farm produce in order to preserve or make them more convenient for consumption. In this study value addition refers to *Fruit Processing*

Common Interest Groups (CIGs): Groups constituted to address issues/problems that affect all the members

1.10 Conceptual Framework

This study sets out to investigate the factors (independent variables) that hinder the sustainability (Dependent variable) of fruit processing enterprises owned by farmer groups. It was hypothesized that the factors that determined success or failure of the enterprises fell into three categories – two primary and one secondary. These categories and their variables are as follows:

Variable category 1: Group Functioning Dynamics (primary)

- ♦ Formation by purpose
- ♦ Leadership/Governance
- ♦ Members committed to Group/ideals
- ♦ Membership benefits

Variable category 2: Intervention/support (primary)

- Capacity Building (Technical skills in production, processing and financial Administration).
- Marketing
- Networking
- Equipment.

Variable category 3: Management Functions (secondary)

- ▪ Production/servicing
- ▪ Processing
- ▪ Marketing
- ▪ Networking

- Financial Administration

Variable category 1 and 2 are considered primary variables that independently or working together to influence the secondary variable in category 3. The relationship between the variables categories and sustainability are shown in figure 1 below.

Figure 1: Researcher's Conceptual Framework

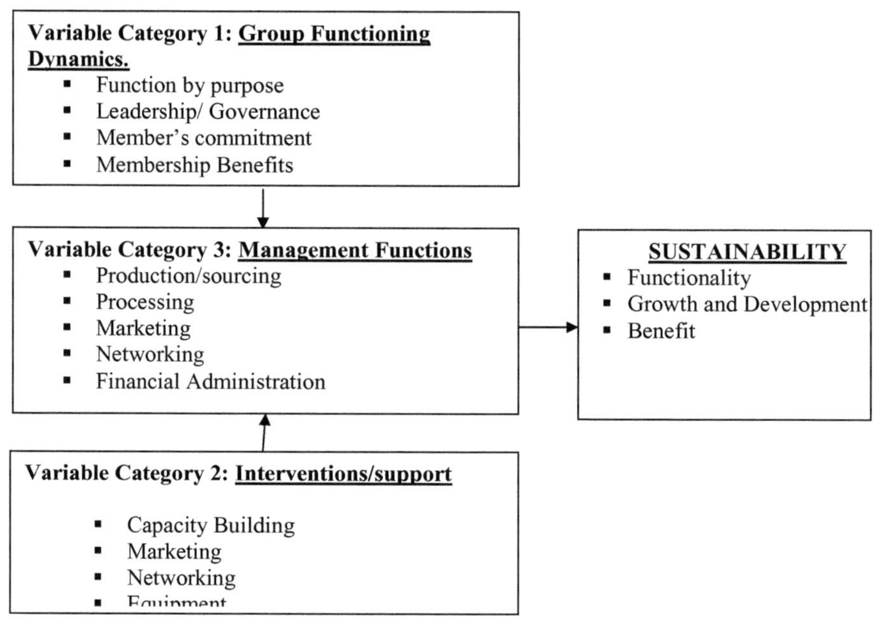

In this hypothesized relationship, sustainability (S) is a function of management (M) which itself is a function of group functioning dynamics (G) and interventions (I). Mathematically this relationship can be expressed as $S=F\ [M\ (G, I)]$

Where: S = the dependent variable – sustainability,

M = the secondary independent variable – management functions

G = the primary independent variable – Group functioning dynamics and

I = the other primary independent variables – Interventions/support.

The rationale for the use of the model was that the purpose for which the groups were initially constituted, the leadership and governance obtaining in the groups, the members' commitment to group objectives, ideals and activities and the real or perceived benefits accruing to members had varying impact on the management functions of the group enterprises. Similarly, the support provided to the groups in the form of capacity building in technical skills for production and processing impact on the management functions. Also support in the form of marketing, networking, financial administration, provision of processing equipment had significant impact on management functions of group enterprise. Consequently, the independent variables determined whether the enterprise would continue to function, achieve growth and development and yield benefits for its members even after being weaned from external support.

CHAPTER 2: LITERATURE REVIEW

This chapter reviews available literature related to theoretical works and research conducted in the areas of: Group dynamics; management of farmer-groups owned small scale enterprises; and interventions, particularly the farmers groups by governments and other development agencies, in Rural Development activities. The review focuses on the achievements and causes of shortcomings of farmers groups owned enterprises in uplifting the livelihoods of the rural poor. It is presented under two subheadings, group dynamics in development activities and interventions and research

2.1 Group Dynamics in development Activities

Although there is ample literature on group formation and functioning, there is little information on why group based enterprises may fail to perform as expected. Since groups are made up of individuals whose interests may not necessarily coincide with those of the groups, it is the contention here that a group will be successful to the extent that the members to are committed to the common interest of the group. Forsyth (2006) defines a group as two or more individuals who are connected to each other by social relationships. Since group members interact and influence each other, groups develop dynamic processes that separate them from a random collection of individuals. These processes include norms, roles, relations, development, need to belong, social influence and effects on behaviours.

On group dynamics Blair (Blair htm) observes that when people work in groups, there are two quite separate issues involved in. These are the task and the problems involved in getting the job done which is often what most people consider. The second issue is the process of group work. This is the mechanism by which the group acts as a unit. Without due attention to their process, the value of the group and even the achievement can be seriously diminished or even destroyed. However, with explicit management of the process, Blair avers that the worth of the group can be enhanced to many times the worth of each individual. This is the synergy which makes groups work attractive in corporate and development settings despite the possible problems in group formation.

To this end, the proposed study concurs with Blair's argument by hypothesizing that both the process and purpose of group formation have a significant impact on the success of group based enterprises. The proposed study will deal with self-help groups registered with the

Ministry of Culture and social services. These groups were formed through three different strategies as follows:

1. The agricultural extension worker selects to work with an already existing self-help group, especially a women's group.
2. The agricultural extension worker encourages farmers to form in order to boost their sales by engaging in value addition to their produce.
3. Farmers are mobilized in large numbers for some specific extension activities e.g. soil and water conservation. The extension work promotes group based farmers' enterprises alongside the larger extension activities.

Groups one and three were initially formed for different purposes from those embraced by farmers' enterprise groups.

In *Group Leadership* (1971) a publication by the institute of Adult Education, University of Dar es Salaam, it is observed that the relations between individuals in a group may affect group work. Regarding group goals it is further noted that groups are born because they aim at something. People get together on realizing that as individuals they cannot solve a given problem or achieve certain goals. Above all, efficient group work depends very much on the group spirit; that is the group members' collective commitment to their goals.

Besides process and purpose of group formation, the other important aspect of group functioning is group leadership. Leadership has to make clear what the aims of the group are. Leadership must never allow any doubts among members. The leader also keeps the members informed and keeps track of progress towards the goals. The leader must strive to strengthen equality between group members. The University of Dar es Salaam cited above posits that the strength of a leader is developed by the group's confidence in him/her. Therefore, the leader is not leading because he/she likes leading but rather because s/he is chosen by comrades to assist them.

According to Blair (html) the leader must constantly be aware that groups are like relationships – one has to work on them. Cleveland (2002), in concurrence with this view, defines 'Team Building' as a process of taking a collection of individuals with different needs, background and expertise and transforming them into an integrated, effective work unit. The other issue associated with leadership is governance. This term cuts across the wide range of leadership styles from purely autocratic to democratic.

Development partners emphasize participatory style of governance where all members have a voice. A leader who promotes this kind of governance is likely to achieve much for the group through higher members' commitment. Hofstede, (1984) in Muchai, 2003; explains the collective orientations as the dimension of culture which indicates the extent to which members of a group or society are closely knit social framework. It is the extent to which people are oriented to the self or to the group. An individualistic culture refer to a loose social frame work in which people are supposed to the care of only themselves and their immediate families, instead of a social framework (collectiveness) in which people are oriented towards group and expect others in the group to look after them. Individualistic culture is characterized by 'I" conscience, self-centeredness, attempt to fulfill obligations to self and has elements of competition, and inclination to private options. Collective culture in contrast is characterized by 'we' conscience, fulfilling obligations to the group rather than to work, considering relationship over task, striving for consensus and a feeling for others.

Lewin (1948) is commonly identified as being the founder of the movement that applies scientific methods in the study of groups. He coined the term 'group dynamism' to describe the way groups and individuals act and react to changing circumstances. Through such studies, Tuckman (1965) proposed the 4 – stage model referred to as the Tuckman's stages for a group. The model states that in group functioning, the ideal group decision making process occurs in four stages: *forming* – pretending to go or get along with others; *storming* – letting down the politeness barriers and trying to get down to issues even if tempers flare up; *norming* – getting used to each other and developing trust; *performing* – working in a group towards a common goal on a highly efficient and cooperative stand.

Tuckman later introduced a fifth stage in the model which he called *adjourning* to refer to the dissolution or breaking up of the group. In reference to the Tuckman Model the study will consider at what stages the groups begun their projects through estimates of the periods of their existence. This is quite significant to the study since some agencies supporting groups consider for how long a group has been in operation while others do not. Some agencies literary form groups through which to channel support to the farmers. Moreover, Kanyiri (2007) the study of Avocado farmer groups' business in Kandara established that the greater number of groups had been in business for more than three years. This period exceeds the incubation period set by the USAID assistance programme to be sustainable. The largest percentage of groups had been in business for more than four years which indicated that the groups were long overdue to be self sustainable.

It has been variously observed that the strongest barriers to project team performance include: communication problems; conflict among team members or between team members and support organization; different outlooks, objectives and priorities perceived by team members; poor qualifications of project/ team leaders; poor trust in and credibility of team leaders; insufficient resources and rewards; lack of project challenges and interest; lack of team commitment; role conflict; unclear team leadership and power struggle and unsustainable project environment.

Conversely, the following conditions are necessary for a team to function efficiently.

- A clear understanding of the project objectives - the revision of project results and benefits accruing from it;
- Clear expectations of each person's role and responsibilities – team members appreciate each other's expertise and contributions and learn how to harness and synergizes these strengths;
- A result orientation – team members have high energy levels, are motivated to go the extra mile to achieve the group objectives;
- A high degree of cooperation and collaboration – members readily share information, ideas and feelings including giving and accepting constructive criticism.
- A high level of trust – team has developed an understanding of interdependence and appreciate the contribution each member make to the team. Team members free to be themselves and openly encourage differences. They have learned how to manage conflict and differences in opinions and view conflicts as opportunities for growth that should not be suppressed.

These views are incorporated in the conceptual framework of this study. They are hypothesized to influence the main functions of the enterprises owned by farmer groups. The absence of any of these characteristics significantly impacts on the sustainability of a groups' agribusiness.

2.2 Interventions and Research

Agriculture has always been the back bone of the Kenyan economy. Currently it contributes 24 per cent directly and 27 per cent indirectly to the country's Gross Domestic Product (GDP). Due to its importance, the GOK has continued to support the sector. For Example the sectoral budget allocation for agriculture grew from Kshs. 15.9 billion in 2003/04 financial year to Kshs 29.7 billion in the 2007/2008 financial year. Such support has been matched with the sector's remarkable growth rate, rising from negative 1.2 per cent in 2000 to the current growth rate of 6.2 per cent (MOA, 2008). The agricultural output at current prices went up by 21.7 per cent from Kshs. 513 billion in 2003 to Kshs. 421 billion in 2005 and stood at about Kshs. 513 billion in 2006 (MOA, 2008:40)

The reported growth has been the result of key reforms and interventions aimed at poverty reduction and wealth creation. Key among the intervention programmes that are being implemented by the ministry of agriculture and that directly affect the farmer groups' agribusiness include:

- National Agricultural Extension Programme (NALEP2) funded by SIDA covering 62 districts.
- Kenya Agricultural Productivity Project (KAPP) founded by the World Bank covering 20 districts;
- Agricultural Sector Research under Kenya Agricultural Research Institute (KARI) and Kenya Agricultural Research Foundation (KARF) funded by EEC/EDF and World Bank;
- Njaa Marufuku Kenya (NMK) covering 71 districts; and
- Rehabilitation of Agricultural Training College (RATC) and strengthening of Agricultural Training Centers (ATCs), formerly the farmers Training centers (FTC) all funded by GoK (MOA, 2008)

The Horticulture Crop Development Authority (HCDA), a state corporation of the MOA, observes that the products grown for local market including fruits have faced a lot of constraints which should be addressed. Key among this is the large quantities that go to waste due to poor market infrastructure and lack of ready market for the perishable products. The authority further proposed interventions which included exploitation of the market potential that exists in the domestic market by streamlining the market systems; improving the market infrastructure and promoting the local consumption; and utilization of horticultural products. Though basic value addition to farm produce was conspicuously left out of the authorities recommendations, it is otherwise included among services it provides to farmer groups which are the focus of this study. The HCDA provides specialized extension services aimed at promoting horticultural production and marketing through organized farmer groups. It registers and inspects fruit tree nurseries to facilitate timely availability of certified planting materials. It also promotes local consumption and processing of horticultural produce (MOA, 2008).

The farmer groups model of production and marketing is considered viable for the Kenyan farmers firstly because the small size of farm holding per individual farmer in most areas. Secondly, because model proved successful before and after independence up to the late 70s with small scale farmers cooperative societies dealing in coffee, tea, cotton and pyrethrum cash crops. Reasons for the collapse of these farmer societies have not been well studied and documented though compared with the groups in this study, the farmers were much larger in membership. However, the principles of group dynamic would apply to them as well.

In their study of women groups in Northern Kenya, Layne, et al, 2005 concluded that the greatest threats to the sustainability of the women groups were posed by external factors. These they identified as draught, scarcity of resources, poverty, and political incitement. Internal factors such as unfavorable group dynamic and illiteracy were also found to greatly impede sustainability. "Principles of good group governance and wisdom in business creation and management were repeatedly stated by respondents as key ingredients for long-term success, making linkages to external development partners is vital to secure access to technology and grants", (Layne, et al, 2005).

The current study hopes to establish among other issues nature of influence that external support to the farmer groups has on the sustainability of the groups and their enterprises. The researcher appreciates the fact that over reliance on external assistance/support may have negative effects on sustainability of the group undertaking by precipitating a dependency syndrome. Mhazo, et al, 2003, on another dimension, state that previous studies have suggested that small-scale food processing activities present a potential source of income for the poorest people in sub-Saharan Africa. However, number of factors may constrain the ability of small-scale enterprises to effectively manufacture and market processed food products. In their cases study of fruits and vegetables in Zimbabwe, Mhazo, et al 2003 concluded that small-scale fruit and vegetable processing are viable provided that appropriate processing equipment, processing skills, including packaging materials and marketing intelligence information are made available. The current study in addition to the above inputs, underscores the importance of group functioning dynamics which has not received much attention in previous studies. This is because for small-scale farmers, the most viable option for them in processing is through self-help groups.

Omiti, J.M, Omolo, J.O and Manyengo, J.U. (2004) in their discussion paper on policy constraints in vegetable marketing in Kenya have proposed several interventions to benefit small scale farmers in the industry which include:

- Use of micro-irrigation technologies to reduce the farmers' over-dependence on rain fed production which currently limits opportunities of small-scale farmers;
- Undertaking research and extension services to improve production, processing, storage and marketing and the introduction diseases and draught resistant crop varieties;
- Formation of grassroots associations for bargaining taxation policies, pricing, transportation, marketing and sub-contracting services;
- Developing and enforcing a suitable code of conduct and standards among the players; and
- Developing market stalls and sheds and other market infrastructure such as processing, cold storage where necessary.

Most of the above recommendations are now the focus of many programmes of GOK's sectoral reforms in agriculture. Their implementation is the responsibility of MOA, its agents the parastatals in conjunction with GOKs development partners. Through another discussion paper of the institute of policy analysis and research (IPAR), Akoten J.E and Ongayo, A.H (2007) explore the determinants of empowerment and food security among small hold farmers in Kenya. These recommend that in order to promote the two-empowerment and food security-the GOK and its development partners should:

- encourage farmers to form associations or groups;
- improve physical infrastructure; and
- Train members in entrepreneurial skills and new production and processing skills.

These proposals apart from another that recommended facilitation of access to credit to the small holder farmers are similar to those presented 3 years earlier by Omiti et al, 2004. On conclusion, it is observed that the process of formulating strategy for small producer groups to grow to small and medium enterprise requires more investigation. This sentiment is shared with Kanyiri (2007) who further noted that, the formation process for small producers groups does not reflect exhaustive analysis for tackling rural poverty. But rather, it is often seems to be personality driver, opportunistic or instinctive approach……" (Kanyiri, 2007:47).

CHAPTER 3: RESEARCH METHODOLOGY

This chapter presents the research design, the population and the sampling process used for the study. It also describes data collection methods - instruments and procedures - and how data were analyzed in order to address the research questions and objectives.

3.1 Research Design

The study adopted a qualitative approach using a descriptive survey design to investigate factors that hinder the sustainability of agri-enterprises owned by small scale farmer groups. The qualitative approach provides for participatory approach which gives a voice to those who are studied (Mugenda and Mugenda, 1999, chambers, 1985). The descriptive survey design was considered appropriate because it allows for a study of a representative sample of the population. Describing the existing situation of a given set of variables, including deductions on basic relationships between them, is the main business of descriptive surveys (Zechmeister and Shaughnesy, 1994:51). Quantitative analysis was conducted to determine the nature of relationship between undertaken variables.

3.2 Population and Sample

Population

The study targeted agri-enterprises owned by small scale farmer-groups which were involved in value addition to fruits. These form the population of enterprises. All ordinary members and officials of these farmer groups, the agricultural extension officers and officials of SACDEP, operating in the two districts - Gatundu and Thika, constituted the population participants for the study.

Sampling Method

The samples for the study were obtained through the use of a multi-stage sampling technique as follows:

Stage 1: A sample of four (4) administrative divisions - Gatanga, Kakuzi, Kamwangi and Thika Municipality - was randomly selected out of a total of seven divisions in the two districts of Thika and Gatundu. Prior knowledge of the area informed the researcher that it was prudent to select a sample of four divisions at the first stage to ensure that there were a sufficient number of operational farmer group-owned enterprises for selection in the second sampling stage.

Stage 2: Two farmer group-owned enterprises were randomly selected from Kakuzi and Kamwangi divisions to form a total of four groups. From Gatanga and Thika municipalities, one farmer group-owned enterprise was selected from each division. This process generated six (6) farmer group-owned enterprises, Mwaka Muki, Ngwataniro, Thayu, Kagaa, Magana Upendo and Thuita that were included in the study.

Stage 3: From each of the farmer owned-group enterprise, two members and two officials were selected on a quota basis to generate a sample of twelve (12) ordinary members and twelve (12) officials. The quota sampling technique was embraced at this stage to ensure gender representation from each of the enterprises and also to ensure that those in the sample are well informed. One extension service provider (MOA extension officer) was selected from each division to form part of the sample. One official from SACDEP was included in the sample. The entire process of sampling provided a sample of 29 participants.

3.3Data Collection

Instruments

Data for the study was collected through individual interviews with members in the three *categories of participants – 24 farmer group members (12 ordinary members and 12 group* officials), four (4) extension service providers and one support organization official from SACDEP. The final interview guides were validated through a review panel consisting of the researcher, the research assistants and two MOA extension officers.

Procedures

The interviews were conducted by two research assistants (RAs) under the direct supervision of the researcher. The research assistants were recruited on the basis of being knowledgeable about farmer group-owned enterprises in the area. The RAs were thoroughly briefed by the researcher who also accompanied them during the initial interview sessions to improve the validity of data that they collected. Relevant documents were reviewed to provide additional data.

3.4 Data Processing and Analysis

Various methods of data analysis and presentation were used to facilitate interpretation of data. Both qualitative and quantitative data analysis techniques were used. In addition, other cartographic methods were employed.

Preliminary data operations entailed processing of data, cleaning and data reduction. Data was coded for easy capturing using computer-based technique, namely; the Statistical Package for Social Scientists (SPSS).

Data analysis was objective based. Quantitative analysis entailed use of descriptive statistics. Cartographic presentation such the use of graphs, pie carts and creation tables have been used to achieve set objectives and afford data greater meaning. This is largely based on what Bailey calls the theoretical principle; driven by the researchers goals and theory (Bailey 1981). Further data analysis entailed subjecting data to statistical tests with the aim of making inference on relationship between data sets or variables.
Regression analysis was used to assess relationships between independent and dependent variables.

A regression function is a mathematical function that describes how the mean of values of a dependant variable changes according to the values of an independent variable. Taking the dependant variable to be y and the independent variable to be x, then for each fixed value of x there is a conditional distribution of y values around the mean (μ) with some standard deviation which is a measure of variability of the y observations that have the same x value.

Linear regression uses a method of Ordinary Least Squares (OLS). This is a method for estimating the linear function that provides the best approximate of the relationship between the interval variables based on observations from a random sample. Alpha(α), betta (β), and the population standard deviation (δ) are treated as unknown parameters and must be estimated in order to estimate the regression equation ($E(Y) = \alpha + \beta x$ for the population. The estimated regression equation can then be used to make predictions about the dependent variable Y and its mean at specific values for the independent variable X.

The following assumptions are made in a regression model:

a) The specification error, that is, the relationship between X_i and Y_i is linear and the dependant and the independent variables are clearly identified.
b) No measurement error: the X_i and Y_i are accurately measured at an interval scale.
c) For the error term:
    - The mean of the error term is zero.

- ❖ The variance of the error component is constant for all values of X_i (homoscedasticity or equal variance)
- ❖ There is no autocorrelation for the error terms for any two observations of *X*. Correlations suggest that there is additional information in the data that has been erroneously omitted.

Correlation Coefficient (*R*)

In determining the association between *Y* and *X*, correlation Coefficient (*R*) and coefficient of determination (R^2) were calculated. The coefficient of determination also provides the goodness of fit for the model. The slope b of the prediction equation $\breve{Y} = a + bx$ indicates the direction of the association between *Y* and *X*. When *b* is positive then there is a positive or upward association meaning that *Y* increases as *X* increases. When *b* is negative the association is otherwise and *Y* decreases as *X* increases. However *b* does not appropriately measure the strength of the association, and thus the need for calculation of correlation coefficient (*R*), which can be described as a standardised slope whose value does not depend on the units of measurement.

Properties of R

a) $-1 \leq R \geq 1$

b) The larger the absolute value of *R*, the stronger the degree of linear association.

c) It is appropriate for use only when a straight line is a reasonable model for the relationship. When there is no linear relationship between *Y* and *X* , then $b = 0$ and $R = 0$

CHAPTER 4: RESEARCH FINDINGS AND DISCUSSION

This chapter presents the data analyzing, findings and discussion organized along the emerging themes that answer the research questions.

4.1 Basic Characteristics of the Farmer Group-owned Enterprises

The farmer group-owned enterprises in this study are defined by the following basic characteristics: the geographical location, duration of existence, group membership and the fruit processing activities.

Table1: Characteristics of farmer group-owned enterprises

	Enterprise	Division	Age	M/ship	Fruits	P/Product	Status	Supporter
1	Mwaka Muki	Kakuzi	10 yrs	26	passion, water melon, mangoes, avocadoes, pineapple	Juices, dried mangoes,	Operates	SACDEP MoA
2	Ngwataniro	Kakuzi	13 yrs	14	Bananas, passion, mangoes, water melon	Juices, dried bananas & mangoes	Operates	SACDEP, MoA
3	Thayu	Kamwangi	3 yrs	28	avocadoes, pineapples, passion	Juices	Operates	MoA, C/Church, Equity Bank
4	Kagaa	Kamwangi	8 yrs	34	avocadoes, pineapples, passion, bananas	Juices, dried bananas,	Collapsed	SACDEP, MoA,
5	Magana Upendo	Thika Municipality	7yrs	30	Mangoes	Juices, Dried mango	Operates	MoA,
6	Thuita Mwihoko	Gatanga	6yrs	14	Passion, avocadoes, bananas	Passion juice, avocado jam	Operates	MoA,

As shown in table 1, all the six farmer group-owned enterprises have been in existence beyond the normal business incubation period of three years. Their membership varies from 14 to 34. It emerged that some of the groups were engaged in other business activities other than fruit processing. Kagaa, which originally had the largest membership, has collapsed as a group enterprise. Consequently, members are operating as individual entrepreneurs. The equipment that was supplied by SACDEP is in use by the former officials of the group. All the groups, except Magana Upendo, grow and process avocadoes and passion fruits. Other fruits grown and processed include mangoes, pineapples, water melon, bananas. All the groups are supported by the Ministry of Agriculture in form of extension services. Three of the six sampled farmer group-owned enterprises - Mwaka Muki, Ngwataniro and Kagaa - receive support from SACDEP.

4.2 Farmer Group-owned Enterprise Leadership

All the groups, except Kagaa, have democratically elected committees to oversee the activities of the group enterprises. The members of Mwaka Muki, Thuita and Magana Upendo rated the leadership of their respective committees as very good. This accounted for the group cohesiveness as indicated by the duration of their existence (at least 6 years), their **Figure 2** below shows how closely rules and regulations are followed in groups

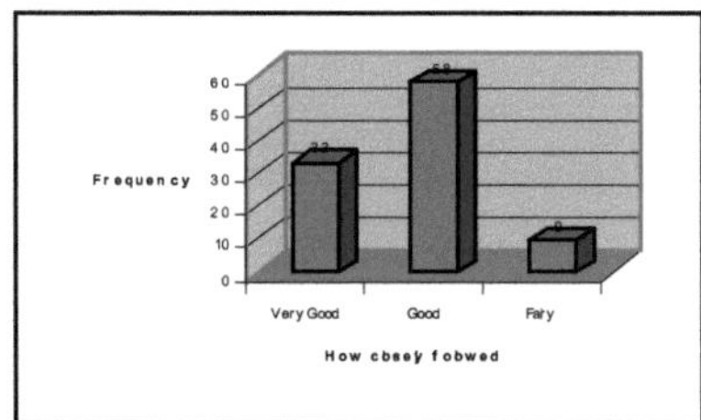

Figure 2: Rules and Regulations

33% of the group members felt that the following of rules and regulations was good and 58% very good. A further 9%, were of the opinion that the rules and regulations were fairy followed.

Figure 3 show the frequency of meetings within groups

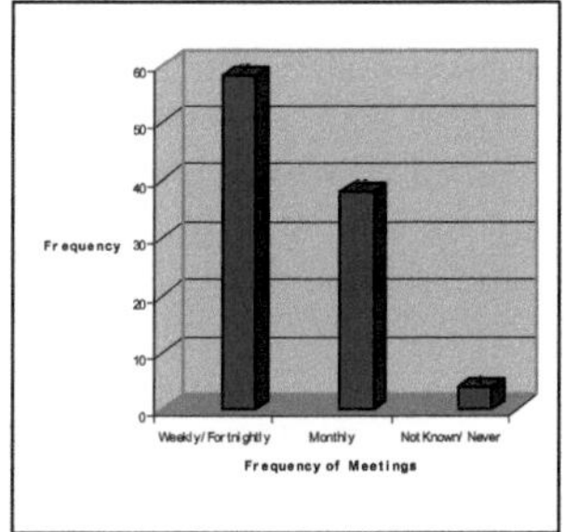

Figure 3: Frequency of Meetings

The findings of the study are that 58%, 38%, and 4% met weekly/fortnightly, monthly and never at all respectively. Frequent meetings are an indicator of a well performing group.

On decision making, 96% of the respondents said that decisions are made by group members. While only 4% said it was leaders/ officials. This shows the groups are democratically led with a lot of member involvement in decision making.

Figure 4: below shows the members rating of satisfaction as group members

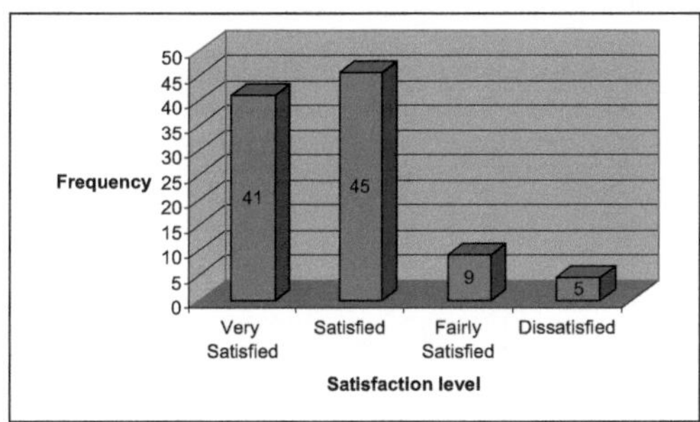

Figure 4: Members' Satisfaction Level

41% of the members were very satisfied, 45% were satisfied, 9% had were fairly satisfied, and another 5% were dissatisfied.

Figure 5 below shows the rating of commitment of members to group

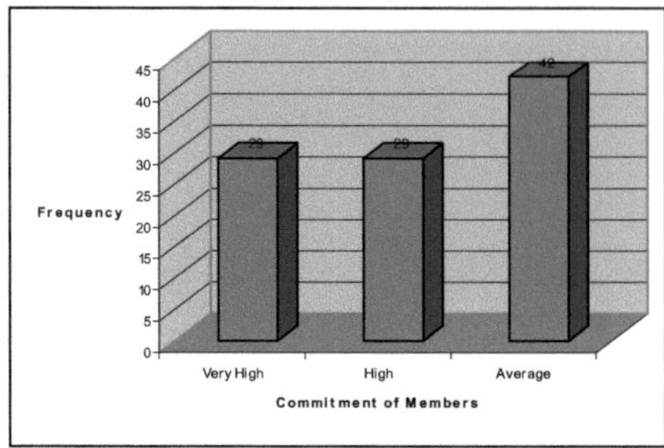

Figure 5: Commitment of Members
Members' commitment was very high for 29%, high for another 29% and average for 42% .

Findings from the study are that 91% had fruit processing among the groups objectives while only 9% did not have fruit processing as a group objective.
Figure 6 shows the benefits group members derived from their respective groups

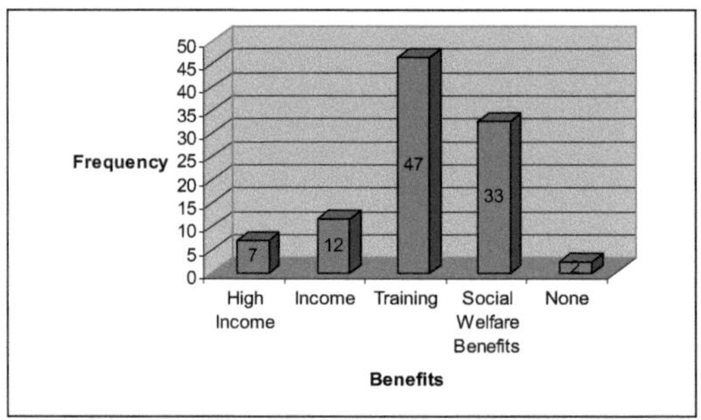

Figure 6: Benefits derived by Members from the Group

The findings indicate that of the group members in the two districts; 7%, 12%, 47%, 33%, and 2% felt that they benefited by way of high income, income, training, social welfare benefits and no benefits at all. The minimal income from the group is a pointer to the fact that groups are generally not yet selling their products in the open markets and therefore, their returns are still low. The question remains why? It is appreciated that although the income levels of the group members are low, they are apparently not so desperate enough to motivate them to fight for the success of their enterprises. It is clear from the research processing still remains a social activity that mainly coincides with the group meetings.

From the findings it is clear that except for Kaaga which already collapsed, all the groups are well led with members who are quite satisfied and ready to contribute to group activities. This means that group dynamics may not be a significant contributor to lack of sustainability of the group projects.

4.3 Factors Affecting the Performance of Farmer Group-owned Enterprise

Fruit Production:

All the respondents reported that their groups experienced shortage in fruit production due to insufficient and unreliable rainfall. This resulted in low volume of fruit processed products such as juices, dried mangoes and jams. Farm inputs were also cited as being very expensive. These included human labour, manure, fertilizer, insecticides and pesticides. Diseases such as flower abortion and insect infestation resulted from inability of farmers to spray the fruit trees owing to the high cost of insecticides and pesticides. The high cost of farm inputs made it difficult for farmers to control pests and diseases. This results in very low yields that are also of low quality. The extension officers interviewed gave pest and disease control as the most limiting factor in the undertaking. Most of the processed fruits except bananas are seasonal. It is therefore not feasible to invest in one type of fruit. This is perhaps the reason why the groups in this study process at least two fruits. The failure to specialize further complicates the groups' ability to produce marketable volume. Hence the reason why they have not explored markets other than their immediate neighbourhood. Seaonality further makes it difficult for the farmers to have regular markets. There are no fruits between the months of May and September, except bananas. Some of the groups have tried to process pineapples which they don't produce but buy from the market. The result is that it has turned out to be expensive given the low volumes of production. Annex 9 gives the production calendar for various fruits in the District.

Fruit Processing:

Some groups process their fruits in the homes of one of their members while the rest operate in rented premises. Many of their processing equipment are not automated; therefore, the production is slow and labour intensive which increases the cost of production. All the respondents cited shortage of fruits as one of the major constraints to fruit processing. They also reported lack of appropriate packaging material and regular machine breakdowns as critical constraints facing their fruit processing enterprises. Some groups reported that their technical staff had inadequate knowledge and skills in modern methods of fruit processing and equipment maintenance.

Table 2 below shows how groups acquired their processing equipment

Source of Equipment	Frequency
Bought by the Group	10
Cost Shared with a Development Agent	31
On Grant	31
Rented	15
Borrowed	5
No Equipment- Manual Methods	8
Total	100

Table 2: How Groups Acquired Processing Equipment

The table above shows a range of sources of equipment used by groups in study area with the majority being cost shared with a development agent and on grant, both rated 31%. Renting and buying by the group members were other significant sources, rated at 15% and 10% respectively. Borrowing was also a source of equipment rated at 5% while 8% of the groups did not have any equipment.

Table 3 shows various places where processing is done

Processing Location	Frequency
Rented Premise	33
Members/ Officials Premise	38
Random (Keeps Changing)	24
Non Members Home	5
Total	100

Table 3: Where processing is done

The findings indicate that most of the processing takes place in either a member's or a group official's premise. This is closely followed by rented premise. The scenario where the venue keeps on changing is significant at 24%. None of the groups own their own premise. This arrangement is not the best for a commercial enterprise. Processing in an individuals premise may not be sustainable in the long run. A venue that keeps on changing may bring about challenges such as failure to meet market requirements and may also create uncertainty among customers. Renting is viable for as long as the processing generates enough money to pay for it. This is possible only if the group does regular processing for some known market.

Figure 7 shows the persons who manage processing.

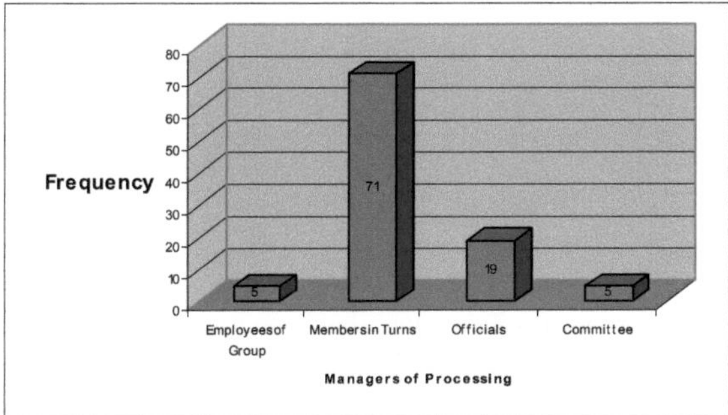

Figure 7: Managers of Processing

Figure 7 shows that 5% were managed by committee and employees of group. A further 71% and 5% are managed by members in turns and officials respectively. Processing is therefore basically done by all the members in turns.

Figure 8 below shows the rating of skill of processing manager

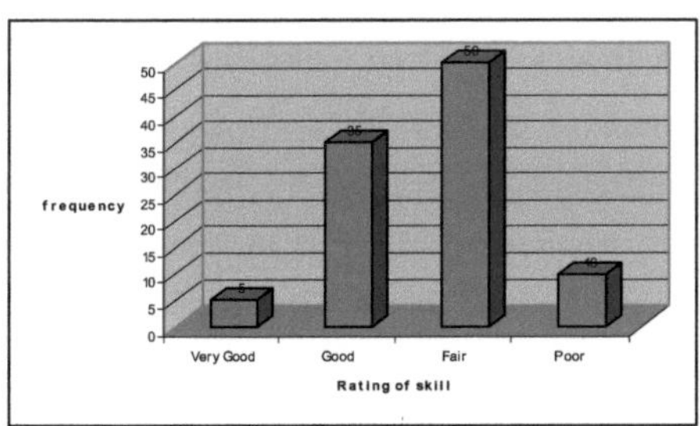

Figure 8: Skill of Processing Manager

From **Figure 8** above, only a paltry 5% of the processing managers had very good skills, 25% and 50% had good and fair skills respectively. Another 10% were poorly equipped with relevant skills. Skills are therefore still a limiting factor to processing and meeting market requirements.

Figure 9 **members rated their fruit processing enterprises.**

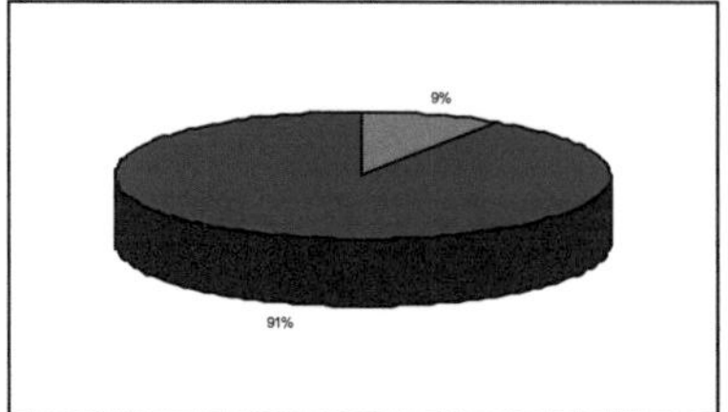

Figure 9: Success of Fruit Processing

As **figure 9** shows 91% and 9% of the group members rated their groups as successful and fairly successful respectively. This is from group members' point of view.

Marketing and Financial Constraints:
Most of the sampled enterprises sell their products in their localities and have no specific customer bases. They all cited lack of transportation and proper storage facilities as some of their most crippling constraints to their business growth and sustainability. Apart from SACDEP, the groups had no other external supporter. Lack of adequate capital and inability to source for capital input are other constraints. Even though SACDEP has assisted Mwaka Muki and Ngwataniro by linking them to internal and external markets, they have not taken advantage of due to lack of transportation and cold storage facilities.

A big percentage of the groups, 57% were marketing their products in the local markets and neighbours. The rest of the groups, 43% were not yet marketing their products. They sell to the members mainly or to occasions. The main bottleneck to marketing has been the inability to package the products and meet the market requirements such as registration of the products. Processing is not done regularly but in all cases, save for Mwaka Muki, it coincides with the season of the fruit and group meetings.

Figure 10 shows whether groups store products before sale

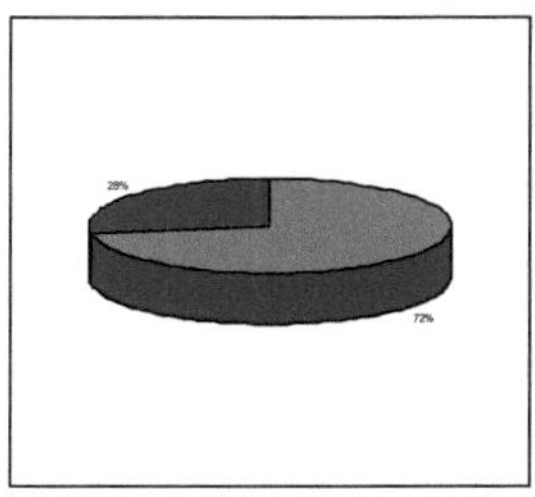

Figure 10: Storage of Products

Figure 10 shows that of the groups in the district; 72% store their processed products before sale while 28% don't store. Stored products are mainly dried fruits.

Figure 11 shows the groups ability to satisfy customers' demands for their products

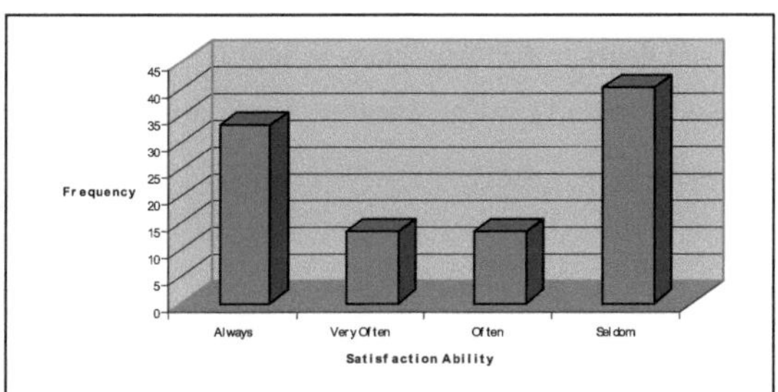

Figure 11: Ability to Satisfy Customers Demands

Figure 11 shows that of the groups in the district; 33%, 13%, 13%, and 40% were able to satisfy demands for their products always, very often, often, and seldom respectively. From this it is clear that the market potential for the processed fruit is high, well above what the farmer groups are able to meet with their current low production levels and substandard and irregular quality. There is therefore room for bigger business in this area.

The market requirements for foods are high. This is right from hygiene to packaging and registration of the product. The groups have problems getting a more or less permanent premise for their operations. Hygiene is also compromised a lot because there is no specific management structure for the operations. As reported, the regular mode of operation is for the members to do it in turns. This does not allow for maintaining any standards of hygiene or food quality. Products sold in the Kenyan market are supposed to be registered by KEBS as a mark of quality.

Figure 12 shows whether groups have any plans to register products with KEBS.

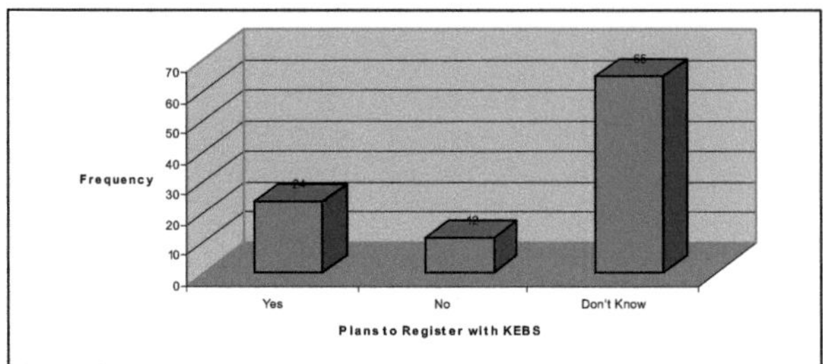

Figure 12: Plans to Register with KEBS

Figure 12 shows that of the groups in the district; 24% had plans to register their products with KEBS while 12% had no such plans. Majority of the members, 65% did not know if there are any plans to register with the body. This clearly shows the groups are very far from meeting the market requirements.

Transportation of products to the market is not done by any particular mode of transport. The regular mode is where buyers who basically direct consumers in this study, collect for themselves. There are few cases where groups get orders from their locality in which case they use means such as vehicles and bicycles to deliver.

4.4 Regression Analysis of Various Factor Affecting Sustainability

In this section regression analysis of various factors involved in sustainability of fruit processing enterprises was done using SPSS.

a) Dependent variable is members' satisfaction while the independent is the rating of enterprise success. The result is as shown below

Model Summary

Model	R	R Square	Adjusted R Square	Std. Error of the Estimate
1	1.000(a)	.999	.999	.825

a Predictors: (Constant), How do you rate your fruit processing enterprise

This means that the relationship between members' satisfaction and success of enterprise is perfect since R is equal to 1.

The regression equation is $Y = 1.550 + 1.43X$ where Y is members' satisfaction and X is the success of the enterprise. The estimation for this relationship is given below. Since R^2 is equal to 0.999 the model is an almost perfect fit for this relationship.

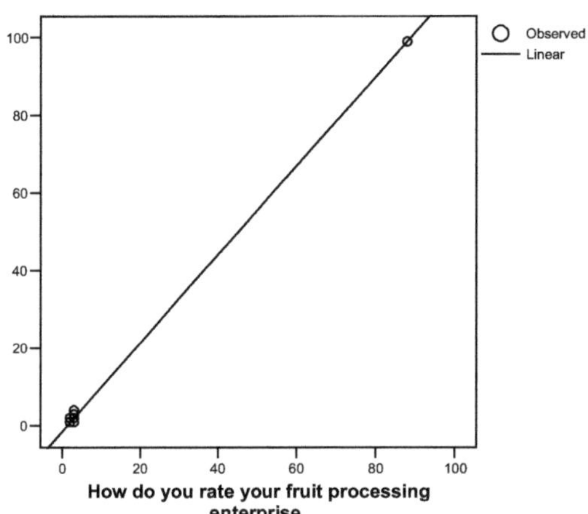

b) Dependent variable is success of enterprise and independent is where processing takes place. The result is as shown below.

Model Summary

Model	R	R Square	Adjusted R Square	Std. Error of the Estimate
1	.793(a)	.628	.611	14.981

a Predictors: (Constant), Where does processing take place

This means the relationship between the success of enterprise and where processing is done is strong and positive. It therefore follows that the success is strongly determined by the suitability of the processing premise/ site.

The regression equation is $Y = 0.275 + 0.671X$ where Y is the success of the enterprise and X is where processing takes place. The estimation for this relationship is given below. The linear model is a moderately good fit for this correlation given R^2 is 0.628.

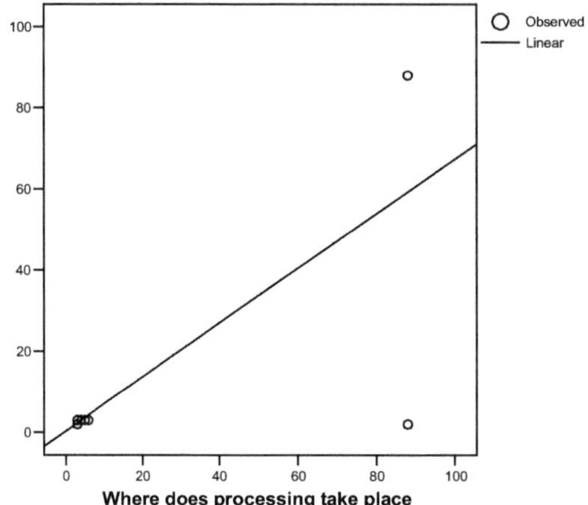

c) Dependent variable is success and independent is how the group acquired the processing equipment. The result is given below.

Model Summary

Model	R	R Square	Adjusted R Square	Std. Error of the Estimate
1	.494(a)	.244	.209	21.362

a Predictors: (Constant), How did the group acquire the processing equipment

The interpretation is that there is a moderately weak positive relationship between success and how the equipment was acquired. It would therefore appear there are other factors that account for success other than how the equipment was acquired.

The regression equation is $Y= 2.94 + 0.301X$ where Y is the success of the enterprise and X how the group acquired processing equipment. The estimation for this relationship is given below. The linear model is a poor fit for this relationship since R^2 is 0.244

The estimation for this relationship is here below:

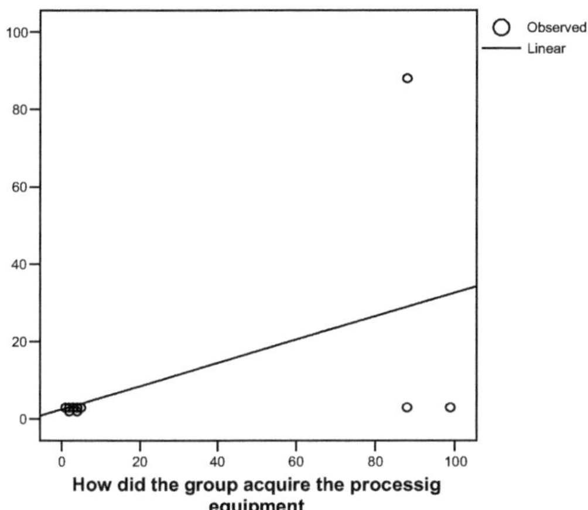

d) Dependent variable is success of enterprise while independent is who manages processing. The regression result is given below.

Model Summary

Model	R	R Square	Adjusted R Square	Std. Error of the Estimate
1	.793(a)	.628	.611	14.977

a Predictors: (Constant), Who manages the processing

This means the relationship between the success of enterprise and who manages is strong and positive. It therefore follows that the success is strongly determined by who manages.

The regression equation is $Y = 1.484 + 0.657X$ where Y is the success of the enterprise and X is who manages processing. The estimation for this relationship is given below. The linear model is a moderately good fit for this correlation given R^2 is 0.628.

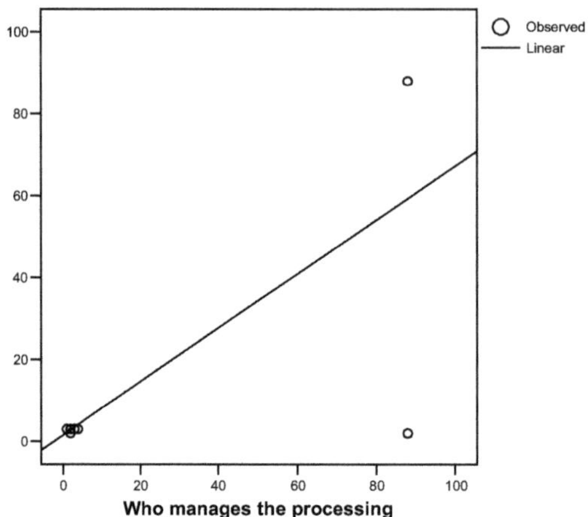

e) Dependent variable is members' willingness to centralize processing and independent is whether fruit processing activity among group objectives

Model Summary

Model	R	R Square	Adjusted R Square	Std. Error of the Estimate
1	.420(a)	.176	.139	33.056

a Predictors: (Constant), Is fruit processing among the groups objectives

The interpretation is that there is a moderately weak positive relationship between members' willingness to centralize processing and whether fruit processing activity among group objectives. It would therefore appear there are other factors that account for willingness to centralize other than group objectives.

The regression equation is $Y= 16.000 + 0.843X$ where Y is members' willingness to centralize processing and X is whether fruit processing activity among group objectives . The estimation for this relationship is given below. This is a poor fit for this correlation since the value of is R^2 0.176.

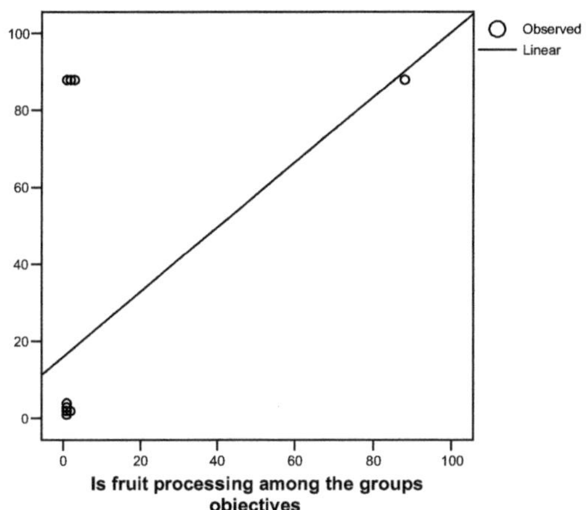

f) Dependent variable is success of enterprise while independent is processing skill of managers

Model Summary

Model	R	R Square	Adjusted R Square	Std. Error of the Estimate
1	.670(a)	.449	.424	18.233

a Predictors: (Constant), Rating of skill of processing managers

There is a moderately strong relationship between success of enterprise and processing skill of managers.

The regression equation is $Y= 1.640 + 0.495X$ where Y is the success of the enterprise and X processing skill of managers. The estimation for this relationship is given below. The model is a moderately poor fit for this correlation given the value of R^2 is 0.449

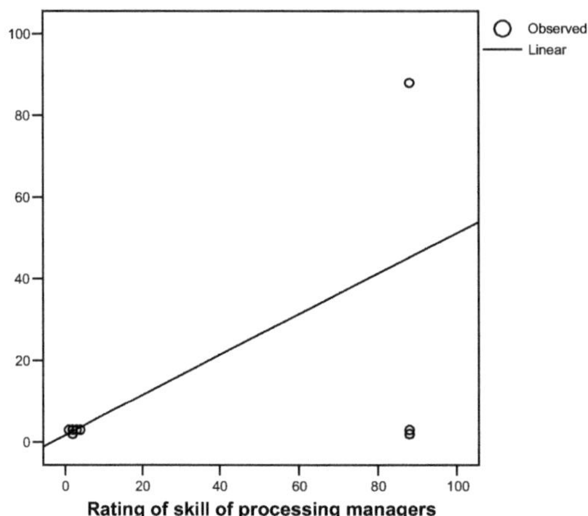

g) Dependent variable is willingness to centralize fruit processing while independent is ability to satisfy customers demands

Model Summary

Model	R	R Square	Adjusted R Square	Std. Error of the Estimate
1	.445(a)	.198	.161	32.622

a Predictors: (Constant), What is your ability to satisfy your customers demands for your products

The interpretation is that there is a moderately weak positive relationship between willingness to centralize fruit processing and ability to satisfy customers' demands. It would therefore appear there are other factors that account for willingness to centralize fruit processing other than ability to satisfy customers' demands.

The regression equation is $Y = 7.160 + 0.349X$ where Y is willingness to centralize fruit processing while X is ability to satisfy customers demands. The estimation for this relationship is given below. The linear model is a poor fit for this correlation since R^2 is 0.198.

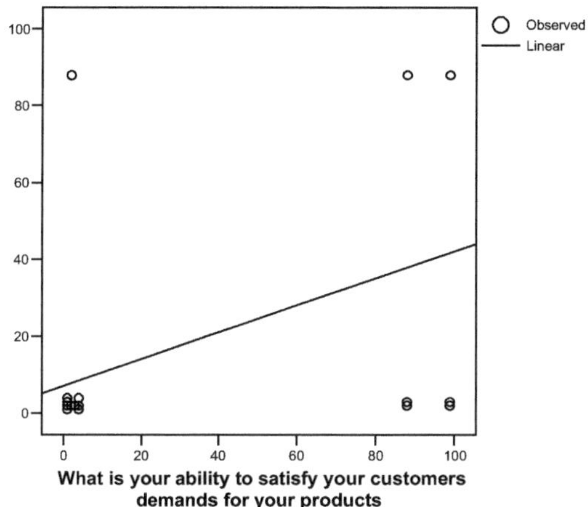

h) Dependent is ability to satisfy customers demands while dependent is the skill of the processing manager

Model Summary

Model	R	R Square	Adjusted R Square	Std. Error of the Estimate
1	.326(a)	.106	.066	43.884

a Predictors: (Constant), Rating of skill of processing managers

There is a moderately weak relationship between ability to satisfy customers' demands and the skill of the processing manager. There are thus other factors that account for customer satisfaction other than the processing skill.

The regression equation is $Y = 29.236 + 0.455X$ where Y ability to satisfy customers demands while dependent is the skill of the processing manager. The estimation for this relationship is given below. The model is a very poor fit for this relationship given the value of is R^2 0.106

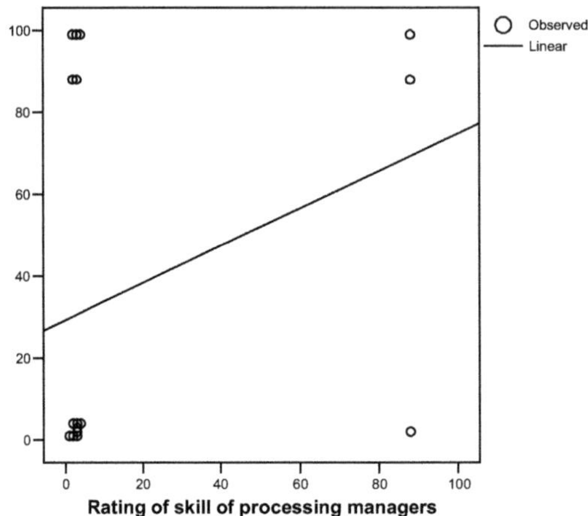

i) Dependent variable is success of processing enterprise while independent is benefits derived from the group.

Model Summary

Model	R	R Square	Adjusted R Square	Std. Error of the Estimate
1	.223(a)	.050	.007	23.946

a Predictors: (Constant), Benefits derived from the group

There is a weak relationship between success of processing enterprise and benefits derived from the group. This means the success is not determined by the benefits the members perceive to get from the group but would be accounted for by other factors.

The regression equation is $Y = 20.685 - 0.135X$ where Y is the success of enterprise and X is benefits derived from the group. The estimation for this relationship is given below. The linear model is a very poor fit for this relationship.

How do you rate your fruit processing enterprise

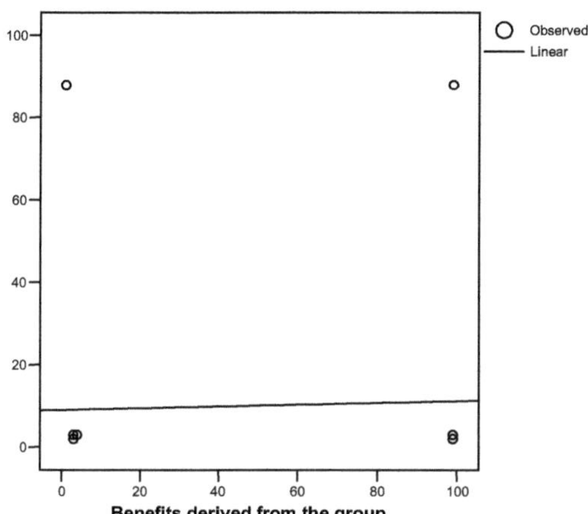

j) Multiple regression function for the success of enterprise as the dependent variable with independent variables as ability to satisfy your customers demands, whether fruit processing among the groups objectives, how the group acquired the processing equipment, members' interest in centralizing fruit processing, benefits derived from the group, who manages the processing, Members satisfaction, Skill of processing managers and Where processing takes place.

Model Summary

Model	R	R Square	Adjusted R Square	Std. Error of the Estimate
1	1.000(a)	.999	.999	.791

a Predictors: (Constant), What is your ability to satisfy your customers demands for your products (C_s), Is fruit processing among the groups objectives (O), How did the group acquire the processing equipment (E), Would member be interested in centralizing fruit processing to involve more groups (C), Benefits derived from the group (B), Who manages the processing (M), Members rating of satisfaction as a group member (S), Rating of skill of processing managers (S_k), Where does processing take place (P)

The interpretation of this model is that there is a perfect relationship between success of the processing enterprise and all the listed variables. This means these variables combined account for success.

The equation for this regression is as follows

$Y= 2.023+ 0.868S- 0.301B- 0.003O + 0.005E + 0.061P - 0.065M - 0.001C + 0.001C_s + 0.005S_k$

Coefficients(a)

Model		Unstandardized Coefficients		Standardized Coefficients		
		B	Std. Error	Beta	t	Sig.
1	(Constant)	2.023	.984		2.055	.059
	Members rating of satisfaction as a group member	.868	.013	.992	68.233	.000
	Benefits derived from the group	-.301	.279	-.010	-1.078	.299
	Is fruit processing among the groups objectives	-.003	.014	-.002	-.199	.845
	How did the group acquire the processig equipment	.005	.006	.009	.940	.363
	Where does processing take place	.061	.203	.073	.303	.766
	Who manages the processing	-.065	.193	-.078	-.336	.742
	Rating of skill of processing managers	.005	.017	.006	.275	.788
	Would member be interested in centralizing fruit processing to involve more groups	-.001	.010	-.002	-.109	.914
	What is your ability to satisfy your customers demands for your products	.001	.005	.001	.147	.885

a Dependent Variable: How do you rate your fruit processing enterprise

Summary of the relationship between variables

The regression analysis above gives the key determinants of success as who manages processing, where the processing takes place and members' satisfaction. The other variables that moderately affect success of enterprise are how the group acquired the processing equipment and the skill of the processing manager. The benefits derived by the members from the group don't seem to have any relationship with the success of the processing enterprise.

4.5 Interventions for Sustainability

All the sampled farmer group-owned enterprises received varied assistance from NGOs and GOK through the Ministry of Agriculture. The extension officers of MOA provided all the groups with capacity building in the areas of crop husbandry and entrepreneurship. Only Magana Upendo group ever received any equipment in the form of solar drier MOA. The Ministry of Culture and Social Services provided seed money of Kshs10,000.00 to Thuita Mwihoko Women Group. KARI also assisted Mwaka Muki with seeds at one time.

SACDEP has provided key interventions to Mwaka Muki, Ngwataniro and Kagaa farmer group-owned enterprises which include technical and entrepreneurial training including sponsored training in institutions of higher learning, equipment such as ripeners, solar driers, juice extractors and Kshs30,000.00 seed money. ATIRI; a programme under KARI, provided training, seeds and fertilizer for one year, drip kits and water storage tank to Magana Upendo group.

Figure 13 below shows the level of members' interest in centralizing fruit processing. This would increase the volume of production and address the constraints faced by the small-scale farmers groups.

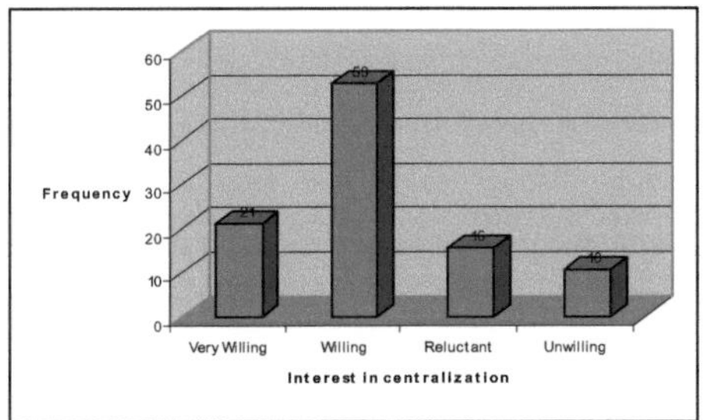

Figure 13: Interest in Centralising

Figure 13 shows that over 70% of the members were willing, very willing and willing i.e. 24% and 54% very high and high. A further 16% and 11% were reluctant and unwilling respectively. This means there exists potential to centralize fruit processing in Thika and Gatundu Districts.

CAPTER 5: SUMMARY OF FINDINGS, CONCLUSIONS AND RECOMMENDATIONS

This chapter presents a summary of findings of the study, conclusions based on the findings and the recommendations resulting from the respondents and drawn by the researcher. The matrix below is considered as the most concise method of that presentation.

5.1 Summary of Findings and Conclusions

Five of the six farmer group-owned enterprises have been in existence for a long time and therefore are likely to continue operating. All the enterprises receive support from the MoA mainly in form of capacity building. The other major supporter of some of the enterprises is SACDEP which provides technical and entrepreneurial skills training and fruit processing equipment. Five groups have embraced a democratic form of leadership comprising well constituted management committees which have ensured group cohesion and continuity. Kagaa's collapse as a group enterprise resulted from lack of democratic leadership. It is further appreciated that support to farmer groups is best channeled through already existing groups. This is as opposed to initiating formation of groups so as to channel support through them.

The study revealed that insufficient and unreliable rainfall, high cost of farm inputs, diseases and pests have reduced fruit production in all the farmer group-owned enterprises, hence low volume of fruit processed products. The study also revealed that fruit processing is constrained by lack of automated processing equipment in some enterprises while in others equipment broke down frequently. Marketing of processed products was mostly restricted to the local consumers because transportation and proper storage facilities were inadequate. Capital inputs and the inability to source for external financial support were found to be crippling constraints to enterprise growth of all the group-owned enterprises. The groups were unable to utilize the internal and external markets that they were linked to by SACDEP.

The farmer group-owned enterprises received major interventions from GoK and SACDEP in form of capacity building and equipment. A few have received small financial assistance in form of seed money and farm inputs.

The farmer groups have not been able to transform themselves into entrepreneurs as would be expected. All the groups in the study except Mwaka muki process the fruits when they have group meetings or when they have occasional orders. This also coincides with when the particular fruit is on season. This is as opposed to regular processing if it were a business enterprise. Secondly, processing is done by members in turns and thus compromising on quality and hygiene. None of the groups have engaged qualified staff to do processing. The members have received training on processing from SACDEP and MOA. However, their skill if still low as rated in this study.

The processing by farmers groups is meant to add value to their own fruits as opposed to buying from the markets. These are seasonal fruits and from the production calendar it is clear only bananas are available throughout the year. This has implications in terms of supplying a regular market. This further challenges the ability of the farmer groups to run this as a business. The option of centralizing and running the processing as a commercial enterprise may ease these constraints because it would be possible to store processed products and at the same time outsource fruits from other areas when out of season in the study area.

The analysis clearly points to the fact that success is highly determined by ***who does processing and their skill, and where it takes place***. The option of centralizing would address these variables because it would put in place a processing plant that meet the market requirements and also have trained technicians to do processing on behalf of the farmers. This would allow farmers to concentrate in fruit production in which they have an advantage and leave trained people to do processing and marketing on their behalf. This would enable the farmer groups to de-link ownership from management which is a key ingredient for success of a business.

5.2 Recommendations

In order to enhance sustainability of farmer group-owned enterprises in Thika and Gatundu districts, the following interventions are suggested:

1. More external assistance in form of capital inputs and further capacity building in crop husbandry and fruit processing need to be extended to the enterprises.
2. Disease and pest control interventions should be put in place to enhance productivity of fruits for all the farmers in the districts.
3. Enterprises should be empowered through organized external market linkages and transportation.
4. Enterprises should be assisted by donor agents and GOK to acquire more up-to-date processing equipment and cold storage facilities.
5. Rural electrification should target the enterprises that do not have power supply so they can use automated fruit processing equipment for more efficient fruit products.
6. To enhance fruit production during dry seasons, irrigation technology should be introduced with the support of GOK and other interested development partners.
7. Centralising of fruit processing would be viable. This would allow for large professionally run enterprises. To overcome the issues of seasonality it would be best to invest in processing of many fruits in the plant as opposed to one. Two plants would be feasible- one to serve Kakuzi and Municipality Divisions and another to serve the Divisions of Kamwangi, Gatanga and Ruiru. These could be either run by a registered farmer entity or an independent entrepreneur. The key to success would be de-linking management from ownership.

5.3 Further Research

Following the insight gained in the course of this study, the following suggestions for further research is made in the area of farmer group-owned enterprises:

1. An in-depth study of group-functioning and management dynamics obtaining in farmer group owned enterprises.
2. A study would be undertaken to establish specific affordable and customer friendly forms of capital sourcing for farmer group-owned enterprises.
3. The viability of farmer group owned fruit processing plant

REFERENCES

Bailey, N. T. (1981). Statistical Methods in Biology. London: Hodder and Stongton

Chambers, R. (1985). Rural Development: Putting the Last First. London: Longman Publishers.

Cleland, I & Ireland, R. (2002). Project Management: Strategic Design and Implementation. Singapore: McGrew-hill Companies.

Community Avocado Producers Groups: A Case of USAID Community Funded Avocado Program in Kandara, Maragua District. A Postgraduate Research Project Report submitted to Institute for Human Resource Development, JKUAT, Kenya.

Coppock D. Layne et al, 2005 (2005). Women's Groups in Arid Northern Kenya: Origins, Governance, and Roles in Poverty Reduction

Forsyth, D.R. (2006). Group Dynamics. 4th Ed. Belmont.

Francis, A. (1990). Business Mathematics and Statistics 1st ELBS Edition; The Guernsey Press Co. Ltd., Guernsey, Channel Islands.

Government of Kenya – MOA (2008). Advertising Feature Article. Public Service Week. Daily Nation, Wed August 20, 2008

Government of Kenya (2007). Statistical Abstract. Central Bureau of Statistics, Ministry of Planning and National Development, Government of Kenya Printers; Nairobi, Kenya.

Government of Kenya – MOA (2006). Strategic Plan for Ministry of Agriculture 2006-2010. Government of Kenya Printers; Nairobi

http://www.see.ed.ac.uk/~gerard/Management/art0.html (retrieved on 16th September, 2008)

Institute of Adult Ed. University OF Dar es Salaam (1971). Group Work Leadership. Tiden – Barnanger Tryckener, Ab 2692 Stockholm, Sweden.

Kanyiri, J.N. (2007). Factors Hindering the Sustainability of SME owned by farmers in Kandara

Kigotho, J.W. (2008). Sustainable Agricultural Magazine. A SACDEP-Kenya Magazine Issue No. 5 April,2008; Thika, Kenya

Lewis, K. (1948). Groups, Experiential Learning and Action Research in http://www.infed.org/thinkers/et-lewin.htm (retrieved on 17th August, 2008)

Mhazo, N., Mlambo, B.,Proctor, S., & Nazare, R. (2003). Constraints to Small – Scale Production and Marketing of Processed Food Products in Zimbabwe: The Case of Fruits and Vegetables. Development Technology Centre, University of Zimbabwe

Moore, D.S. (1995). The Basic Practice of Statistics. W.H. Freeman and Company; New York, USA.

Muchai, A.M. (2003). Management Styles and Organization Culture: Implications for Entrepreneur Development in Kenya. M.A. Thesis

Mugenda, O.M. & Mugenda, A.G. (1999). Research Methods: Quantitative and Qualitative Approaches. ACTS Press; Nairobi, Kenya.

Musembi, O (2008). Mangoes from above Rescue Farmers in Semi-Arid Regions. Daily Nation Newspaper, Finance Magazines: Thursday, August 28, 2008: Nairobi, Kenya.

Omiti, J.M., Omolo, J.O. & Manyengo, J.U. (2004). Policy Constraints in Vegetable Marketing in Kenya. Discussion Paper No. 061/2004: Institute of Policy Analysis and Research (IPAR) Discussion Paper Series; Nairobi, Kenya

Onyango, A.H. & Akoten, J. (2007). Determinants of Empowerment and Food Security among Smallholder Farmers in Kenya. Discussion Paper No. 097/2007: Institute of Policy Analysis and Research (IPAR) Discussion Paper Series; Nairobi, Kenya.

Tuckman, B. (1965). Development Sequence in Small Groups. Psychological Bulletin 63:384-99. Naval Medical Research Institute, Bethesda, MD

UNDP (2006). Growing Sustainable Business: Fruit Processing Project Report. UNDP: Kenya Country office. Nairobi.

Zechmeister, E.B. & Shaughnesy, J.J. (1994). A practical Introduction to Research Methods in Psychology. 2nd Ed. McGraw-Hill, Inc.

ANNEXES

Annex1: Farmers'/Group Leaders' Questionnaire

Part 1: Preliminary Information

1. Name of group:
2. Name of respondent:
3. Location of the group:
4. Year formed:
5. Position in the group:

Part 2: Group leadership and other dynamics

1. Does your group have rules and regulations (bylaws or constitution)?

   i. Yes ii. No iii. I don't know
2. If the answer to (1) above is yes, how closely are they followed?

   i. Very good ii. Good iii. Fairy iv. Poorly v. very poorly
3. What is the frequency of group meetings?

   i. Weekly /Fortnightly ii. Monthly iii. Quarterly iv. Annual
   v. not known/ never
4. Who makes decisions for the group?

   i. members ii. Leaders/ officials iii. Chairman/ chairperson
5. How do you rate your satisfaction as a member of the group?

   i. Very satisfied ii. Satisfied iii. Fairly satisfied iv. Dissatisfied
   v. Very dissatisfied
6. What benefits do you derive from the group?

   i. High income ii. Income iii. Training iv. Social welfare benefits
   v. None

7. How do you rate the commitment of members to the group?

   i. very high ii. High iii. Average iv. Low v. Very low
8. Is fruit processing among the group's objectives?

   i. Yes ii. No iii. I don't know
9. How do you rate your fruit processing enterprise?

   i. very successful ii. Successful iii. Fairly successful
   iv. Unsuccessful v. Failed

Part 3: Production and processing

1. Production and processing statistics

Types of fruits process ed	Area / No	Produc tion in tons	Cost of producti on- gross margin	Months of product ion	Proces sed produc ts	Quantity of each for the group	Cost of processing/ ton	Price of finished products/ unit of sale

2. How did the group acquire the processing equipments?

   i. Bought by the group ii. Cost shared with a development agent (s)
   iii. On grant
3. Where does processing take place?

   i. Group's premise ii. Public/ community facility iii. Rented premise
   iv. A member's / official's premise v. Random (keeps changing)
4. Who manages the processing?

   i. Employee(s) of group ii. Members in turns iii. Officials
5. How do you rate the processing skill of the response in (3) above

   i. Very good ii. Good iii. Fair iv. Poor v. very poor
6. Would you be interested in centralizing the fruit processing activities to involve more groups, say, at divisional or district headquarters?

   i. Very willing ii. Willing iii. Reluctant iv. Unwilling v. Not sure

Part 4: Marketing

1. Where do you sell your produce?

2. What is the distance to the market in Kilometers?

3. How do you transport your produce to the market?

   i. By vehicle ii. By bicycle iii. By hand iv. Through agent
   v. Collected by buyer
4. Do you store the processed products before sale?

   i. Yes ii. No iii. Don't know
5. What is your ability to satisfy your customers' demands for your products?

i. Always ii. Very often iii. Often iv. Seldom v. Never

6. Does the group have any plans to register the product(s) with Kenya Bureau of Standards (KEBS)? i. Yes ii. No iii. Don't know

Annex 2: Extension Officers' Questionnaire

Part 1: preliminary information

1. Name of officer:

2. Division:

3. List Groups in the division doing value addition of fruits
   a.
   b.
   c.
   d.
   e.
   f.
   g.

Part 2: Production and processing of fruits in the division

1. production and processing statistics for the division

Types of fruits processed	Area/ (ha) No	Production in tons	Cost of production- gross margin	Months of production	Processed products	Quantity of each for the group/division	Cost of processing/ton	Price of finished products/unit of sale

2. The constraints to fruit production in the division are
   i. Pest and disease control ii. Fertilizers/manure iii. Knowledge of fruit farming iv. Inadequate rainfall/ irrigation v. Poor planting material

3. Please rank the constraints you chose in (2) above from the most significant to the least significant
   i.
   ii.

iii.
iv.
v.

Part3: Marketing

1. Do you think centralizing processing would be more viable for the groups?
   i. Yes ii. No iii. Not sure
2. If answer to (1) above is yes does the pooled production from the groups amount to enough for a bigger plant?
   ii. Yes ii. No iii. Not sure
3. If the answer to (1) is No please give reasons
   a.
   b.
   c.
   d.
   e.
4. Please give the challenges encountered by groups in getting their products registered by Kenya Bureau of Standards (KEBS)
   a.
   b.
   c.
   d.
   e.
5. To what extent would pooling ease these constraints?
   i. Fully ii. Partly Not at all
6. Is it possible for the groups to make profits from the local market?
   i. Yes ii. No iii. Not sure

Annex 3: Farmers'/Group Leaders' Interview Guide

1. Name of group:
2. Location of the group:
3. Year formed:
4. Is your group performing well?
5. Please describe its achievements.
6. Type of fruit(s) grown (total for the group)
7. Acreage fruit(s) grown: (Total for the group)
8. Briefly describe the type of ownership of premise where fruits are processed:

a. Rented

b. Group owned

c. Community/public owned

9. Please describe the leadership of your group (i.e. management committee, etc)
10. Please explain how the processing is managed.
    a. By individual members
    b. By employees of the group
    c. Describe the sources of the fruits which are processed.
11. Describe how your fruit products are marketed.
12. Please describe who the main consumers/customers for the processed products

13. Who provides/provided the technical skills capacity building in processing/production.

14. Who provides/provided entrepreneurial skills you need to succeed in this business?

15. What support do you get from the government donors/NGOs?

16. What benefits do you get
    a. From the venture?
    b. From belonging to the group?

17. Do you get any financial assistance and if so from whom?

18. What are the achievements of the group?

19. What are the challenges you experience in relation to:
    a. Production of fruits/sourcing
    b. Processing
    c. Marketing
    d. Transportation
    e. Group organization
    f. Financial support
    g. Technical skills needed for processing

20. Please give at least four recommendations that can help you do better

Annex 4: Extension Officers Interview Guide

1. Name of officer:
2. Division:
3. List Groups in the division supported by SACDEP
4. Please comment on the cooperation between extension officers & SACDEP & other NGOS involved in agri-businesses.
5. Please describe briefly the type(s) of support/service provided by Ministry of Agriculture
6. Please comment on the performance of the groups with respect to:
    a. Organization
    b. Production/sourcing
    c. Processing and packaging
    d. Marketing
    e. Profitability
7. Describe briefly challenges faced by:
    a. Farmers
    b. Extension officer
8. Recommendations

Annex 5: SACDEP Interview Guide

1. Please list the groups that you have supported in Thika for food processing ventures:

   A.

   B

   C

   D

   E

2. Please describe briefly the type of support your organization has given to the following groups:

GROUP	DESCRIPTION OF TYPE OF SUPPORT PROVIDED		YEAR & DURATION	REMARKS
	Production /Sourcing			
	Equipment			
	Processing			
	Capacity Building			
	Marketing			
	Networking			
	Others			
GROUP	DESCRIPTION OF TYPE OF SUPPORT PROVIDED		YEAR & DURATION	REMARKS
	Production /Sourcing			
	Equipment			
	Processing			
	Capacity Building			
	Marketing			
	Networking			
	Others			
GROUP	DESCRIPTION OF TYPE OF SUPPORT PROVIDED		YEAR & DURATION	REMARKS
	Production /Sourcing			
	Equipment			
	Processing			
	Capacity Building			
	Marketing			
	Networking			
	Others			
GROUP	DESCRIPTION OF TYPE OF SUPPORT PROVIDED		YEAR & DURATION	REMARKS
	Production /Sourcing			
	Equipment			
	Processing			
	Capacity			

	Building			
	Marketing			
	Networking			
	Others			
GROUP	DESCRIPTION OF TYPE OF SUPPORT PROVIDED		YEAR & DURATION	REMARKS
	Production /Sourcing			
	Equipment			
	Processing			
	Capacity Building			
	Marketing			
	Networking			
	Others			
GROUP	DESCRIPTION OF TYPE OF SUPPORT PROVIDED		YEAR & DURATION	REMARKS
	Production /Sourcing			
	Equipment			
	Processing			
	Capacity Building			
	Marketing			
	Networking			
	Others			

3. Please describe your achievements in your efforts to support these groups.

4. Please describe the challenges that you face in providing support to these groups.

5. Please make recommendations for making these groups sustainable

Thank you so much and God bless you.

Annex 6: Production Calendar for Various Fruits in Thika District

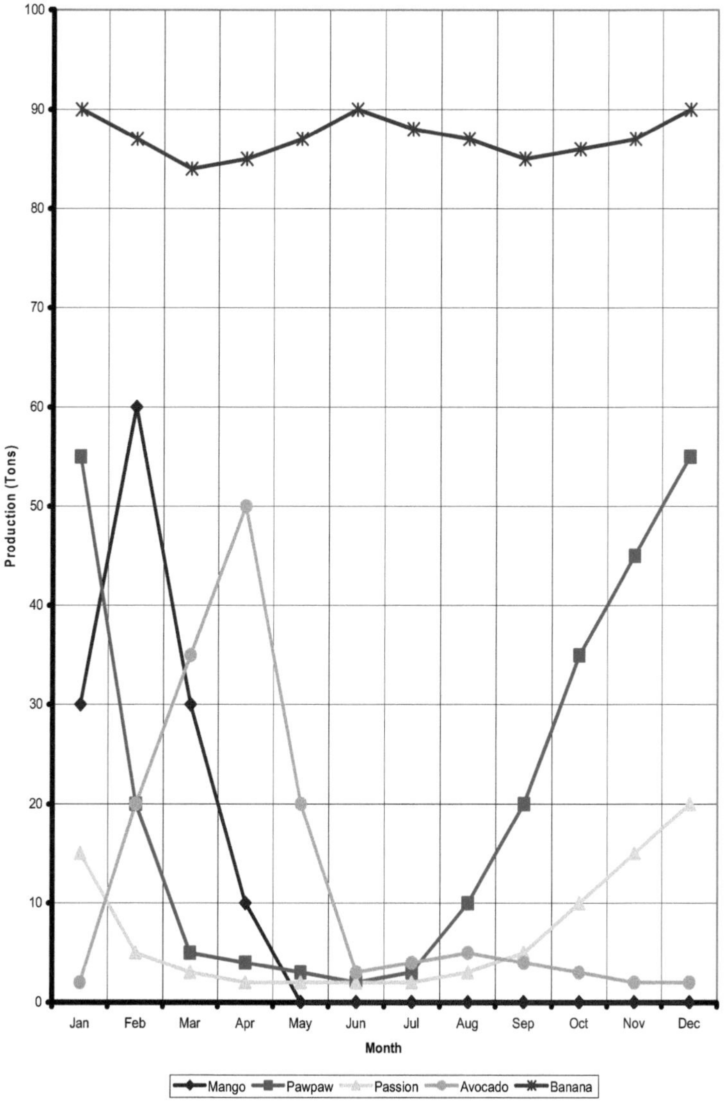

Annex 7: Map of Kenya Showing Location of Thika District

Annex 8: Map of Central Province Showing Location of Thika District

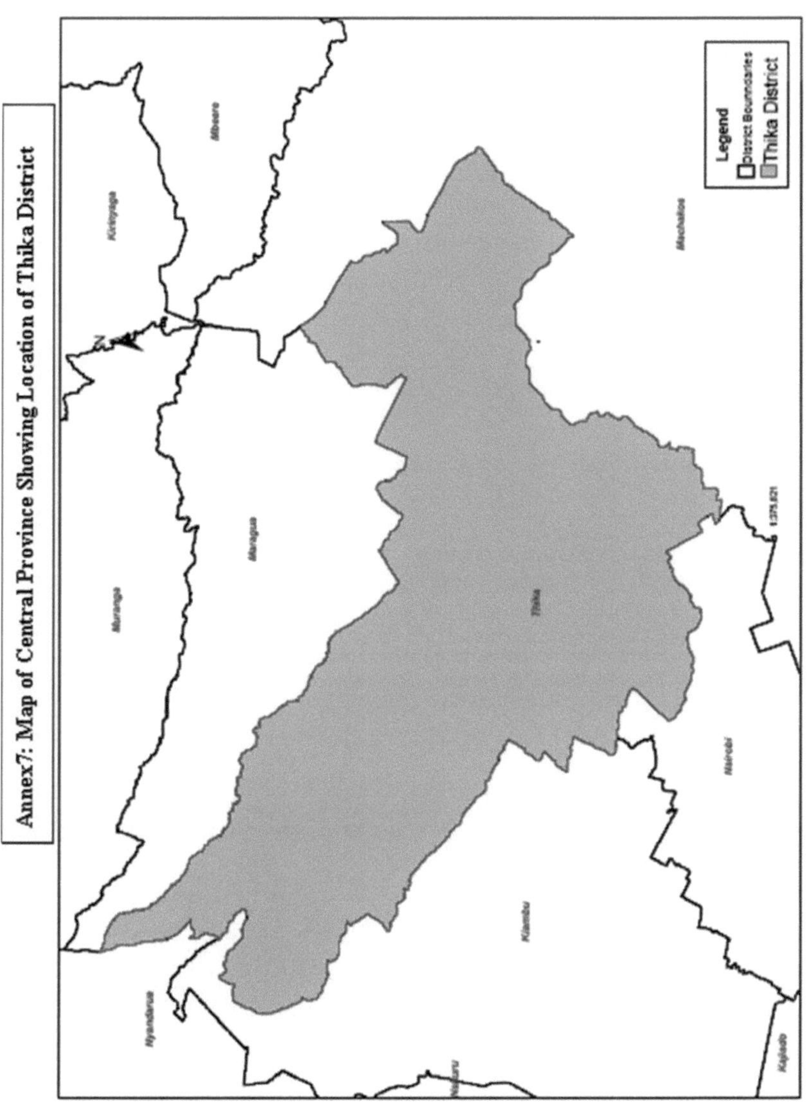

Annex 9: Map of Thika showing location of groups

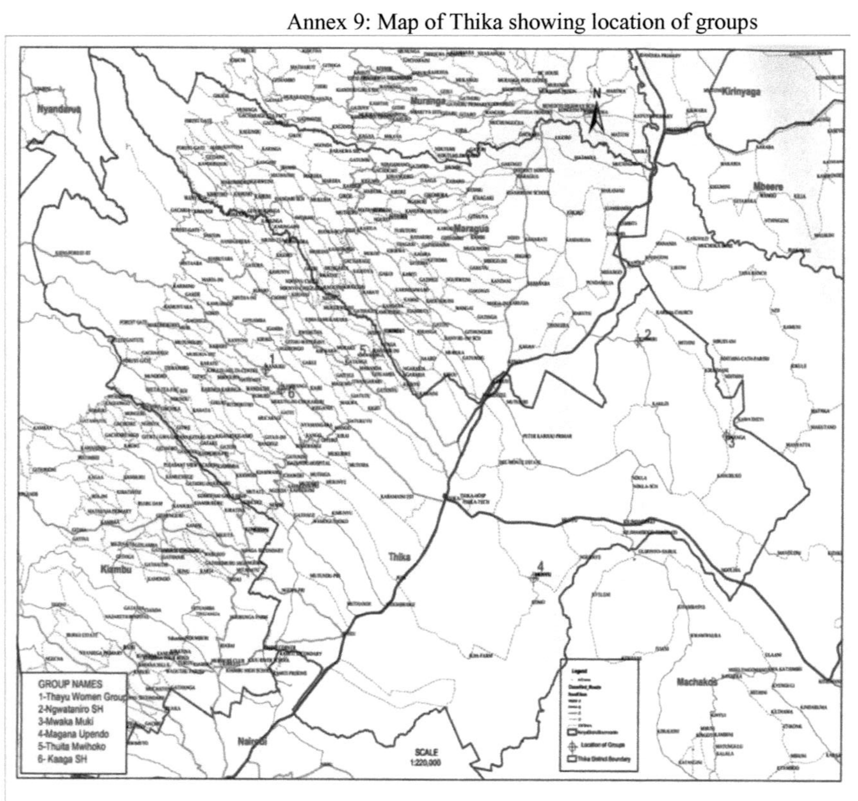

Printed by Books on Demand GmbH, Norderstedt / Germany